E=mc^8

ENERGY by INTENTION
(THE PRIMER)

$E=MC^8$

ENERGY
BY
INTENTION
(THE PRIMER)

WITH ALBERTA EINSTEIN,
ENERGY EXPLORER

"Words matter…Energetic intention matters more."

Published February 2021

©2020 Laurie Bonser writing as Alberta Einstein

Book design by Pamela Trush, Delaney-Designs.com

Printed in the United States of America

Library of Congress Cataloging-in-Publication Data

Bonser, Laurie, 1962 –

E=mc^8; Energy by Intention (The Primer) 2021

Print ISBN   978-0-9904142-4-7

eBook ISBN   978-0-9904142-5-4

"Believe in
your own craziness...
because that's where
your greatness comes from."

Alberta Einstein, January 2020

$E=mc^8$

ENERGY by INTENTION

With Alberta Einstein, Energy Explorer

Alberta Einstein is the intentionally adventurous/humorous pen name for an individual who delights in pondering, discovering, questioning and exploring all parts of life. on this glorious planet of Earth and in the endless Universe ~ fully embracing human life experiences in its ever-shifting expansion and cosmic experiments.

I invite all readers to open wide the throttle of your imaginations, explore and embrace each experience and interaction, and discover ever more personally unique joys and contributions in your own individual experiment of Life.

Regardless of your education, profession, passions, or philosophies, my intention is that the key factors contained in $E=mc^8$ will provide a supportive catalyst to assist you further in unlocking, developing, and sharing your own special creations.

"Whether you can observe a thing or not depends on the theory which you use. It is the theory which decides what can be observed." Albert Einstein 1926

FOREWORD

Serapis Bey, 2017

"At a time when change is progressing more rapidly and with less linear form than ever before in the history of human-kind, _we offer this material to Simplify, De-mystify, and Support the "being' part of action_. It is increasingly important to fully realize the need for each individual to STOP the chaotic madness the human world has defined as 'Living' and instead enter into a thoughtful, mindful consideration of building the meaning-filled life you are intended to experience.

Time is past to dwell on measuring karma, insisting on simplified pre-packaged creation theories, and debating which tagged methodology or product or approach is THE newly found answer to success, happiness, love, and every other enticement known to marketing and promotion.

Scientists in so many fields now verify through physical and visual means the ability of world matter to transform instantly, through distance ~ and into innumerable forms and outcomes that support the human experience in ways beneficial and miraculous and startling, and yet so flowingly logical.

The Universal Laws must be more visible, experienced, and prominent than ever before on Earth – even the visible daily light tone, hue, and brightness is a regular driver toward change and higher understandings. Are you aware and paying attention to these signs?

More individuals than ever before are receiving and sharing previously more selectively 'channeled' books. But be still and hear your *own inner voice* and encouragement to 'jump on the time wagon' and enjoy the new ride of hope, love, manifesting, trusting, expanding, being, and discovery.

Always listen to and trust your own intuition, and respect your own subsequent choices, above ANY and ALL other sources.

Cries of poor me, too much to do, what else can I do, they made me do it, not my fault, unlucky in life, that's the way it is, etc. are falling on deaf ears above, dear ones. When you ask and open to assistance, you'll have more support than you could conceivably imagine. Without your initiative, however, there's compassion and energy on stand-by here, but nothing directly that will be done to 'fix' your situation or future.

There is equally a time and place for definition, action, and movement. Through Grace there is an amazing opportunity for immediate release, change, alignment, and creation available to you now as never before. Study the past, the teachings, the protected messages, the cherished tools from a variety of honored teachers and peoples.

But understand also that there has been a great condensation of time, space, opportunity, and energy in the past years that take these venerable approaches and expedite their application at quantum x 10 compared to linear timelines."

Purpose of this Primer

The Energetic Intention here is to encourage our understanding and joyous exploration of <u>Personal Quantum Energy (PQE)</u> in a physical world:

- ❀ Personally recognize, acknowledge, and apply the essence of this Energy ($E=mc^8$) available to each of us

- ❀ Tap into existing Universal Energy streams and *create* new ones from a conscious, mindful, proactive position

- ❀ Effectively stimulate *intentional* physical manifestations, holistic personal healings, scientific and philosophical discoveries and expansion, universally-beneficial global systems, supportive and integrated economic interactions, meaningful environmental stewardship, and so much more

- ❀ Better grasp the concept of how much more powerful <u>Personal Quantum Energy</u> is than the physical/visual system we usually elevate to 'illusion reality' status

- ❀ Experience how ALL perspectives and offerings tie together to reinforce the energetic Universal laws and responses

- ❀ Recognize that the old paradigm of duality (choices limited to either this or that) cannot be sustained with any credibility or success

❀ Support the concept of 'zero degrees of separation' (connections and entanglements)

❀ Highlight the importance and finesse involved in creating the 'energy boiling point' – the seemingly smallest tweaks that take us from simmering to manifesting

❀ **Above All Else:** *approach and use this study solely in the universal spirit of love, integrity, consciousness, highest good, respect, beginner mind openness, and infinite impact.*

This primer is definitely NOT a scientific or mathematical treatise, a spiritual channeling, a religious philosophy, or any claim to present a nicely buttoned-up explanation of the Universe or Life or Energy or any other theories.

It is an open starting point for greater human awareness, empowerment, and action on a considerably larger scale than we've enthusiastically and practically embraced in quite some time here.

So, here's to the art of increasing finesse and integrity while learning, experimenting and creating!

Setting the Stage

Before we take off into the wild world of this new/old equation theory and its current relevance, we need to cover three very important foundations for this entire discussion so everyone can stay focused.

First, some introductory comments for the variety of readers here ~

For those of you who may primarily self-identify as scientists, mathematicians, engineers, Western medical professionals, technology gurus, and other STEM (science, technology, engineering, math) focuses:

This primer does not contain white paper research, conference-style data backup, blind trial test results, or any other similar attempted proof of particular positions or theories to convince anyone or hold forth on a 'truth'.

Its purpose, instead, is to highlight and expand on those universal qualities, laws, and experiences (seen and unseen) that can meaningfully add to your future research, computations, educational outreach, and personal understandings.

Please take and use whatever following information and theories that spark a resonance with you, permit your intuition to be open to some alternative language and concepts...and thereby continue to be a future creator and enlightener in your own unique field and mission.

"The most beautiful experiences we can have is the mysterious. It is the fundamental emotion that stands at the cradle of true art and true science. Whoever does not know it and can no longer wonder, no longer marvel, is as good as dead and his eyes are closed."
Albert Einstein 1935

For those of you who may primarily self-identify as artists, healing arts practitioners, spiritual coaches and supporters, alternative and complimentary medicine professionals, writers and communicators, liberal arts teachers, and the multitude of related focuses:

While this primer does include references to science and mathematical concepts for educational and structure purposes, the overarching purpose is to reinforce how all areas of study can combine and support one another in the consistent universal laws, qualities and experiences.

So please allow your intuition to absorb whatever information is relevant to you at the particular time you are reading this, and set aside any nagging internal thoughts that technical information and details are beyond your comprehension or present 'need to know' scope. With this peaceful mindset you will continue to expand your unique offerings in meaningful and incredible ways.

"Do the difficult things while they are easy and do the great things while they are small. A journey of a thousand miles must begin with a single step." Lao Tzu

---

Second, the wisdom of the ancients carried forward a set of timeless principles against which both science and art are to be considered...and these Principles are core to the testing and summary to all experiments, experiences, and understandings.

They are the keys to the ideas offered throughout this primer, and all are vital components to your future applications. You may wish to bookmark this spot and refer back frequently for reference and inspiration - or perhaps even pause here an acquire your own copy, as of the study of these concepts can easily span lifetimes.

The Hermetic Principles – Fundamental Foundations

"The principles of Truth are Seven; he who knows these possesses the Magic Keys before whose touch all the Doors of the Temple fly open." The Kybalion

1. *The Principle of Mentalism*: The All is Mind, the Universe is Mental

2. *The Principle of Correspondence*: As Above, So Below; As Below, So Above

3. *The Principle of Vibration*: Nothings rests; everything moves; everything vibrates

4. *The Principle of Polarity*: Opposites are identical in nature, but different in degree; Extremes meet; All truths are but half-truths; All paradoxes may be reconciled

5. *The Principle of Rhythm:* Everything flows, out and in; Rhythm compensates; All things rise and fall

6. *The Principle of Cause & Effect:* Every Cause has an Effect; Every Effect has a Cause; Chance is but a name for Law not recognized

7. *The Principle of Gender:* Gender is in Everything; Everything has its Masculine and Feminine; Gender manifests on all Planes

Third, bear in mind throughout all the following chapters and ensuing information, the main message of this primer can be succinctly summed up with the enduring phrase *"May the Force be With You... and In You"*

Note: The 'Resource and References List' at the back of the book provides excellent background and foundational information on the various subject matters (both scientific and metaphysical) mentioned in the following chapters.

Personal Quantum Energy

"You may not feel outstandingly robust, but if you are an average-sized adult you will contain within your modest frame no less than 7 X 10^18 joules of potential energy—enough to explode with the force of thirty very large hydrogen bombs, *__assuming you knew how to liberate it and really wished to make a point.__*"
Bill Bryson, *A Short History of Nearly Everything* 2003

Ok, stop and read that quote above again please. After your initial quick laugh, you may have been ready to keep reading and see what 'Alberta' has up next. But when you really sit with this information and the entire idea, even for a few seconds – or a minute or two or three - there's an incredible sense of wonder, astonishment, uncertainty, hugeness, opportunity, 'oh geez', 'oh my gosh', 'oh bleep, what do I do with this now' that begins to develop.

It's so huge for most of us that it's really tempting to set it aside and dismiss all the implications...because it now opens the door further into a world of creating, power applied with Universal intention, and our own personal roles and contributions. This is not simply an opportunity for physical action and responses, but a call out to <u>transform</u> the mind/ body/soul connection and <u>integrate</u> unlimited personal energy to <u>create</u> healthy life experiments and experiences that truly benefit us at our highest sense.

The whole point of this primer is to help us acclimate to the scale of our own personal potential, to affirm that this is indeed our birthright and history and future, and invest thoughtful contemplation and focus into practice and exploration and application.

It's not a venture for the faint of heart, for quick remedies, an out-of-context sound bite, or a new gadget or short-term fix 'just because we can'. So, let's hang in there together and be brave; this is the long game folks.

Here's some overlying concepts for the following paragraph:
- ✓ ***Energy precedes Form***
- ✓ ***Energy is liberated Matter***
- ✓ ***Matter is Energy waiting to happen***

Readers with science backgrounds can no doubt outline the various physical elements, stages, progressions, identifications of molecules and building blocks, and latest discoveries from the labs and experiments. Hold onto those foundations and visualizations as we move forward...and prepare to spend even more time with the concept that VISUAL observation and documentation *<u>does not</u>* define or own the existence of Energy itself.

Readers with initially perhaps a more conceptual vision or 'intuition' of Energy and its impacts can likewise rely on your own personal experiences and connections...while at the same time rejoicing in the physical acknowledgements, signs, and symbols which provide additional affirmation to humans on a physical plane that energy manifestations are indeed subject to the Universal Laws which express consistently and in a universally understandable foundation.

Universal Expressions of Quantum Energy

My personal favorite question is "Why?" That is closely followed by "How?" I also like "Trust and Verify", and the old family saying "Love Many, Trust a Few, and Always Paddle Your Own Canoe".

So, I'm always delighted each day to see and feel how the Universal principles and foundations manifest consistently in so many different expressions, both in physical and metaphysical forms. Consider the following few examples... some familiar to you based on your past studies and work, and others that will yet further compliment your understandings and discovery:

- ❀ Sacred Geometry & Patterns
- ❀ Numerology
- ❀ Nature Patterns and Sequences
- ❀ Human body interaction with frequencies Hz (tuning forks, drums, singing bowls, all instruments and voice)

❀ Astronomy: stories, maths, cycles

❀ Planetary consciousness interchanges

❀ Plant and animal spirits

❀ Therapies with light, color, water, air, music

❀ Physical forms comprised of water, hydrogen, carbon, stardust elements

❀ DNA structures, gene cross-consistencies/ relationships

❀ Multi-dimensional building blocks

❀ Cross dimension communications

The most interesting follow-up question
for each of us then:

How will I, as a unique individual soul, acknowledge and express both my own special perspectives and service contributions during life on Earth now...and far beyond...with Personal Quantum Energy?

Moving from Reacting into Creating

There are many insightful and helpful writings available from authors/teachers/coaches who explain very well how to recognize our human tendencies to <u>react</u> to outside events, people, and stimulus in ways that aren't always most productive or helpful to us longer term. They provide examples of these situations and how we can choose to respond differently in a more conscious manner. I often hear people say 'you can't change what you feel but you can change how you react'…and I agree this is an important and much more healthy perspective than the straight 'action/reaction' default often experienced.

In our discussions here, though, I propose that we are indeed most capable and deserving to <u>create</u> lives in which the actual existence or appearance of unhealthy or unwanted feelings are very rare to begin with… and accordingly we can then spend even more of our precious and momentous personal energy on reaching higher into discoveries, expansion, and integration of all life elements.

For example: Instruction and practice in the first 'conscious reaction' view definitely gives us more options to choose from in situations with family members, colleagues, other drivers, people we interact with in public settings, etc. to include perhaps more patience, more thoughtful replies, healthier boundary setting, keeping situations in perspective, opening to different opinions and concerns, and much more.

Beyond these steps, though, study and practice in the 'creating outcomes' focus critically adds on our abilities and energetic power to generate and influence life in a very proactive, 'set the stage', 'plan for the desired results' type of focus.

Of course, in order to *create outcomes*, we need to develop a very comprehensive, well-considered mastery of our own personal intentions, skills, and understanding of how we fit in with the Universal picture....in other words, the dedicated and ongoing application of our own personal homework with as much honesty, openness, and receptivity as we can possibly tap at any given point in time. As I mentioned before, this goal is a long-view infinite in essence....and while seemingly sometimes overwhelmingly frustrating it is also overwhelmingly rewarding and inspiring and fun!

The ongoing contemplation and selecting of our intentions – whether for purposes of work, research, service, relation-ships, self-development, policy determinations, or other – really must be the first step and highest consideration when in a creating mode. Taking the time to understand our mo-tivations, goals, our capacity for being open to options and collaborations and feedback, drawing out the blocks and shadows and healing those parts within...these are the steps that cannot be avoided or rushed or judged along the journey to increasing mastery and finesse.

It's far too easy for us to allow situations or projects to take on a 'life of their own' and it's not until much further along that we either develop or acknowledge the troubling realization that the outcome we're experiencing is very far from our original, higher purpose intention.

Continual monitoring, checking-in, priority testing, and vigilance is required to stay the course of masterful alchemy. As we'll discuss a bit later on, the idea of balance (Yin and Yang) is one constant, challenging, exciting component of all this self-study and practice.

Regardless of the applications, the adages of 'Know Thyself' and 'Seventh Generation' are extraordinarily important for each of us in doing our part to maintain the goals of highest and best purposes. Earth is an ultimate learning platform for all of us, and as such there is some Universal built-in leniency and flexibility to accommodate this experiment and experiences.

However, those Universal Principles noted early on in this primer are the fundamental operational keys - and we fail to give them appropriate respect and attention at our own risk for harmful and painful outcomes.

"The key to growth is the introduction of higher dimensions of consciousness into our awareness."
Lao Tzu

So Why E=mc⁸?

The first version of this primer was to be based on the title of Energy by Intention: $E=mc^5$...relating to our current human progression on Earth/Gaia from a third dimension (physical criteria of height x width x depth) through the fourth dimension (adding time) into a fifth dimensional space (adding simultaneous experience/entanglement).

Whether or not a human being is in a place to understand or accept multiple dimensions, it's key to remember that we can still inhabit perspectives and responses in all realms on a choice-basis...and it's very rarely in as clean or clear or absolute a space/level as we would like to categorize!

Examples of **Third** Dimension human perspectives:

- ❈ A person sees themselves as a separate entity in comparison to other people, nature, and the Universe
- ❈ Life is made up of 'duality,': there are good and bad decisions/thoughts/people/situations
- ❈ Life has a finite set of physical benefits or rewards, so a person must to compete with others in a win/lose to survive and/or get ahead
- ❈ A person's role and responses are defined by expectations inherent in labels and titles: family status, job/work, certification, award, age, culture, etc.

❀ Outside forces determine a person's life in terms of happiness, comfort, security, rewards, opportunities, experiences, and more

❀ There is a ranking system of value for all people and a method of judging who fits where on the scale

Tangible expressions examples:

➢ Sell products and services that make money (a substitute for personal security and value) regardless of 'collateral damage'

➢ 'Me' versus 'Them/the World' perspective

➢ 'Labeling' and 'judgment' abounds for roles, religions, politics, movements, groups, tribes, genders, etc.

➢ Financial resources allocated based on influencers and money brokers

➢ Short term solutions and quick fixes

➢ Medical/pharmacy largely based on premise that they human body is flawed and can never be healthy without external professional intervention. There is a pre-determined physical track and outcome that applies to everyone.

➢ Use of Earth resources without respect

➢ Use of fear and division to manipulate and control others

Examples of **Fourth** Dimension human perspectives:

- ❀ The human attention shifts to *include* the pursuit of knowledge and understanding
- ❀ The internal human ego becomes an enemy and a goal is to suppress or eliminate it in an effort to counteract harmful impacts of power, control, and manipulation
- ❀ Labels and judgments/expectations still exist but now also include spiritual pursuits, roles, and accomplishments
- ❀ A person wants to take more initiative over their personal life and outcomes however it's still often a 'me' versus 'others' approach whether in respect to other humans, governments, cultures, businesses, philosophies, religions, etc.
- ❀ Personal influence or impact is recognized but seen as far less powerful than outside forces

Tangible expressions examples:

- ➤ Offer products and services with the direct or implied message of 'this is more enlightened, more nurturing, better for you, aligned with morality, etc.'
- ➤ 'Us' versus 'Them' perspectives and language gather attention
- ➤ The concepts and application of 'discernment' and 'observation' begins to rise in important above 'labels and judgment'

> ➤ Financial resources and use disputed between 'good and bad' morals and applications

> ➤ Short term programs and quick fixes still prevail, but expand to include spiritual/expert/truth solutions

> ➤ Medical/pharmacy based largely on external symptom relief and intervention rather than personal inner knowledge and power

> ➤ Use of earth resources with consideration of recycling, reusing, repurposing

> ➤ Use of fear and uncertainty in ways which more subtly sway people to a particular message or agenda

Examples of **Fifth** Dimension human perspectives:

> ❀ A person begins to realize that darkness/shadow is not an external force to battle against but rather a reflection of internal characteristics, choices, and options

> ❀ The human ego is no longer simply bad or good, but contains elements of both benefit and harm depending on how it is used

> ❀ Healing oneself and becoming whole is the key to both personal expansion and helping others (people, Earth, animals, Universe)

> ❀ Being willing and able to relate to life outside oneself with respect, empathy, and communication is desired and nurtured

❀ All forms of life are constantly changing, cycling, re-evaluating, creating…and as such each person becomes more comfortable and secure with a greater level of uncertainty and adaptation

❀ There is a decrease in 'judgment' and an increase in observation and discretion in choices

❀ The concept of god/source/creator/universe/ energy begins to be recognized as being within the human form, and no longer as existing solely as an outside or external form

Tangible expressions examples:

➢ Offer products and services with the intention to empower each individual to think and choose consciously what is appropriate for their situation

➢ 'WE are Humanity' because of and despite our perceived differences

➢ Money and finances are viewed as neutral resources to be used in support of conscious priorities and intentions, on both the individual and society levels

➢ Longer term perspectives (i.e., seventh-generations considerations) take hold in increasingly more fields and industries

➢ Medical/pharmacy regains more focus and education on human synchronization with Earth/ nature and self-knowing/healing power (versus external cures of all types)

➢ Use of earth resources with very conscious consideration of amounts used/tapped, long term impacts, clearing of past damage and assaults, and regular expressions of gratitude and appreciation

➢ Conscious communications develop which respect diverse perspectives and allow for varied expressions in a balanced and secure environment

I'm sure readers will recognize these various stages and cycling experiences in both their own lives and in societies on a larger scale... so the part(s) that each of you choose to play right now reflects immediately within your own personal life journey AND contributes to how the rest of the world shifts as a whole.

You'd think all that was more than enough to presume to offer material going from E=mc² to E=mc⁵, right?! Well...

After several months of alternating between frustration, confusion, and impatience while writing, I eventually realized we needed to have an even higher, more comprehensive goal to consider so we didn't all get stuck part way through our metamorphosis and end up with a partial accomplishment... rather like a caterpillar who attempts to emerge from a cocoon but only the head and a front leg or two actually makes it into the action of freedom.

Hence the final title of *Energy by Intention: E=mc⁸*, the "8" and "∞" which encompass the concepts of infinity, abundance, unlimited exploration, never-ending opportunities and discovery, continual cycling through all and full recognition that utilizing and balancing all aspects of mind, body, and soul are vital and attainable.

Whew! Quite a large subject to address, but it feels well worth the efforts to stretch and open even further to reach the very best creative fruits available to all of us. We're in the midst of an incredible convergence of energies and connections right now, so let's go for the top waves and head for the stars......

Physics Original Theory 1905:
E=mc²

"...To work in the service of life and the living... In search of the answers to questions unknown... To be part of the movement, part of the growing, part of beginning to understand..." John Denver, _Calypso_ (Recorded 1975 BMG Music)

In this spirit of continuing curiosity, dialogue, experimenting, and a sense of mutual respect and purpose across the ages, we now move further into the high-level concepts of energy, mass, and implications with this very basic summary:

The $E=mc^2$ equation reflects Energy as equal to Mass x the Speed of Light squared and in this basic form refers fundamentally to matter <u>at rest</u> (i.e., that is in a stationary state or non-moving to begin with). As the speed of light squared is a huge number when applied to a stationary mass (448,900,000,000,000,000 in units of miles per hour) the amount of energy bound up in even the smallest mass boggles the human mind....in Alberta's language, simply _ginormous_! And if you move into Einstein's more elaborate kinematics equation which takes into account matter already in motion, you have an *incredibly ginormous* amount of energy occurring.

The equation also presents Mass (matter) and energy as interchangeable, and grew out of work on special relativity –a subset of the theory of relativity which is often times considered to be his greatest overall achievement in the scientific arena. *So, in a nutshell, the legacy of E=mc² is as a way to understand the most basic natural processes of the Universe.*

The applications of these factors to date focus primarily on physical manifestations in science, technology, space exploration, medical analysis, industry, defense, such as:

- ❈ PET scans and diagnostics
- ❈ Smoke detectors to exit signs
- ❈ Radiocarbon dating
- ❈ Power for space travel and telecommunication satellites
- ❈ Neutrino (subatomic particles) detector
- ❈ Nuclear energy
- ❈ Propulsion engines of varying power sources
- ❈ Battery driven operation of devices and vehicles

The forms taken by energy and utilized in physical creations are as liquids, solids, gases, and plasmas. Energy also exists as vibration, light, sound, color, and the elements under Universal law.

Some forms can be visually identified, yet many others still remain beyond current human visual eye sight. For example, the sun and other stars shine as their interior atoms (mass) fuse together, creating the tremendous energy and outward reach that we experience from an extraordinary distance here on Earth. This is also the same basis of energy interactions such as Reiki, Shamanism, The Matrix, and many more titled expressions.

Some industrial/manufacturing developments are deemed 'helpful' to humanity in terms of 'progress' and 'easing burdens' but are often very light on the consideration of long-term impacts and whether compatible with full human potential and desired/sustainable Life/Universal Laws. For example, consider the ongoing philosophical debates and questioning regarding the development and use of atomic bombs and other related applications.

In addition, there are many products developed and sold 'just because we can' with little or no consideration of a full product lifecycle: what happens to the materials once an object no longer functions, sometimes in as few as months or a few years? The result has been land fill waste, chemical leaching, depletion of natural resources, pollution impacts on humans and nature, medical impacts of previously touted 'acceptable levels' of exposure to matter/waste/by-products...and 'out of sight' SPACE Trash!

It is critical that the heart and intention portion of creating being expanded and developed in the human approach, or we will continue to fail to address the core foundations of our assumptions and outputs. Eventually we find ourselves reaping little or no true benefits and instead experiencing an increased detrimental backlash of outcomes across all spectrums including environmental, educational, physical/mental/emotional health, relationships, resource utilization, communications, governing, and soul purposes.

Now is the time to take the incredible life contributions of Albert Einstein and add in the additional finesse factors that will launch us successfully into managing and creating Personal Quantum Energy...in better service to humankind, the Earth, and the Universe.

Personal Quantum Energy Theory 2020:
$E=mc^8$

"The key goal is to use Personal Quantum Energy to create desired results within the fully supported foundational Universal Law Hermetic principles **and** fully respect the highest gift of free will choice of each individual soul." Alberta Einstein 2020

Why the Personal Quantum Energy discussion now? As mentioned in the prior section, people have demonstrated unequivocally that only tapping into a portion of the creative energy focus leads to long term damage, ineffectiveness, side-effects and impacts, destroyed societies and cultures, continual war cycles, harm to Earth/atmosphere/space and beyond.

Each of us can think of examples we've observed, or that honestly enacted ourselves, where short cuts have been taken to reach an end outcome – and experienced the ramifications that come back as boomerangs to clip us again, and again, and again.

We have well rationalized blind spots or areas of denial and negligence... knowing we are omitting heart or mind or spiritual or environmental or physical considerations in the push to 'get it done' or 'move on' or 'get the reward'. And we have equally well rationalized post-outcome excuses and blame placement after the fact as an attempt to soothe our egos and mitigate personal responsibility.

However, with conscious choices and free will responsibility there can, and will, be different outcomes – specifically CREATIONS that we truly support, embrace, and affirm our highest perspectives.

And now for the Equation components...

E=   Energy by Intention

M=   Intention/mass

C=   Focus x Direction x Speed

        x Magnitude x Elements

        x Integration x Universal Law Cohesion

        x Universal Connections

The intentional energetic legacy of $E=mc^8$:

Stimulate, integrate, and cohese the CREATIONS of Personal Quantum Energy

Energy (E) = the expressed creation by directed intention and input factors

Mass/Intention (M) = the substance of matter, concept, desire, purpose, or motivation

Factors (C) = The identified characteristics which impact Energy in varying outcomes based on selection, application, and degree of finesse (next section chart details)

The Finesse Factors and Layers Chart

As you read through the following chart, consider the concept of fusion in all its various descriptions, including: zero points decision matrix, spiral convergence codex, DNA frequencies, the Sophia Code, the Matrix Architecture, building blocks of atoms, etc. Keep in mind also that extra one degree finesse factor which allows for reaching a full desired impact of your intention, such as the transition from 211 degrees Fahrenheit to the 212 degrees Fahrenheit boiling point.

Physicists view *consciousness as a fundamental element*, the actual human brain being a separate smaller entity in perspective that processes information. It's vital to make a sincere effort through this consciousness, to include all factors up front - from first tries through each iteration of increasing finesse- for truly meaningful and significant contributions.

Adjusting the combination and balance across the entire spectrums of components is what determines the outcomes and the legacies of experimentation.

The following is the first edition chart of factors and layers...
my intention is that the detail, examples, and honing of
application will expand much further as we all continue to
experiment and share suggestions:

Finesse Factors	Finesse Layers
Intention/Mass	**D**efinition of thought **M**otivating reasons **S**ingular/composite **C**ohesive/consistent **F**ree will choice respect
Focus	**C**larity of purpose **I**ntention mindset **D**egree of receptivity **D**uration/repetition **L**evel of preparation
Direction	**S**traight-line, Spiral, Wave **B**ubble/cloud **C**hannel/shape **E**xpansion/contraction **C**lockwise/counter clockwise **P**ast, present, future (SpaceTime)

Speed	Variable/constant
	Accelerating/decelerating/gravity
	Density of matter
	Color of Light/Frequency
	Resistance/friction/flow
	Linear/non-linear (SpaceTime)
	Impact of 'spaghettification'
Magnitude	Scope of impact desired
	Energy application needed
	Strength of energy source
	Targeted/defined effect
	Partnering/collaboration
Elements	Heat/cold/variation/levels
	Physical materials to use and in what form (liquid, solid, gas, plasma)
	Elemental spirits to incorporate
	Intrinsic material properties
	Combination effect/interactions
	Understanding permissions
	Honoring purposes
	Locational principles and context: North, South, East, West, Above, Below

Integration	**P**ersonal Energy Balance **S**elf-clarity/character **R**econciliation: ideas, feelings, and emotions **L**ifetimes experiences **U**niversal knowledge access and application
Universal Law Cohesion	**E**xtent of incorporating the Hermetic Principles in creations **V**ibration **R**hythm **P**olarity **G**ender **I**ntegration **C**ause and Effect **Z**ero-point choices
Universal Connections	**T**apping into higher consciousness in all forms, guidance, and information. **S**trength in numbers and knowledge **T**eam work makes the Dream work

An important reminder for integrating all resources and perspectives comes from the frustration and confusion that arises when individuals rely on just one portion of the equation to fulfill an entire outcome, namely that of focusing primarily on affirmations, symbols, mudras, theories, postulations, pros/cons, and the like.

Generally, a partial focus fails through one of the following:

1. Repetition of any type of statement (mind based) without the actual complimentary passion of commitment (heart based)

2. Having a burning desire for an outcome (heart based) without the accompanying action steps and personal accountability (mind based)

3. Moving to take action without having made a conscious commitment (heart based) AND without having prepared to analyze or focus on the outcome potentials (mind based)

Words matter...
Energetic intention matters more.

The End of Duality Flashpoint

Harmonizing all *seemingly* opposing energies

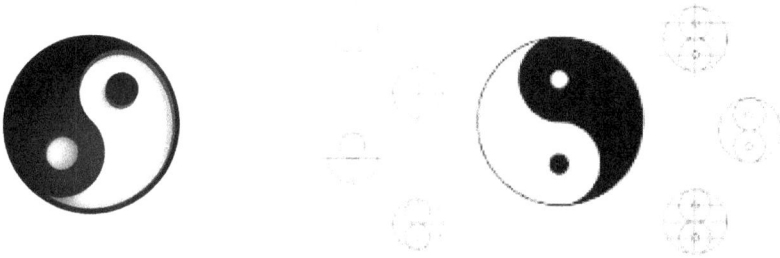

The creative composition of the Yin/Yang symbol illustrates the beauty and consistency of sacred geometry, and holds a universe of stories and meanings as well.

One such often overlooked message is that within each of the solid color shapes there is always a portion of the other color as a counterpoint...within each and every perspective or approach there is also the potential for a different or opposing viewpoint or outcome to be thoughtfully explored and considered.

Within the illusion of duality or opposition is the opportunity to meld together a complimentary, inclusive, highest purpose result – IF we take the time and space to allow discussion, musing, theorizing, and limits removal. It is only by entering into life conversations and creating in this spirit of open possibilities that we can truly listen, discuss, collaborate, create, and continually adjust our conscious choices and contributions.

➢ How can we assist someone but not overstep into enabling/codependency/personal choice?

➢ How can we produce a new product that doesn't pile up in a landfill as permanent waste?

➢ How can we create a new device that benefits human life without replacing awareness and involvement?

➢ How can we maintain community/state/country without perpetuating an endless destructive control-based cycle of action and re-action?

➢ How can we meld seemingly diverse governing policies yet not fall to a lowest common denominator settlement outcome?

➢ How can we sustain and nourish our physical bodies while respecting Earth and her resources?

➢ How can we reach for the new, the challenge, the better solution without steamrolling over others in fear, disrespect, lack of abundance, or undeveloped self-esteem?

Throughout our lives there are many such areas where there are not easy issues with quick solutions, but they are most critical to undertake...sometimes in smaller portions to avoid being overwhelmed and giving up...sometimes in larger leaps of recognition and collaboration than previously expected.

There is an honored time and place for stories. It is important to explore yourself, share messages with others, contemplate what has been learned from experiences, and to note a baseline for re-writing those portions you wish to change.

However, if you remain only in story telling mode (or more likely, copious amounts of re-telling) then the resulting reiteration, reinforcing, embellishing, and regurgitation really just reinforces the old linear habit of explaining and justifying oneself according to whatever label or theory or explanation that can be *oh-so-carefully and reasonably stated.*

And nothing is more disingenuous, false, or insidious than misusing Energy in the name of a human-defined 'Truth' to support a Fear-based ego.

The Entanglement Component

In July of 2019, a team of physicists from the University of Glasgow unveiled the first-ever photo of quantum entanglement in the Journal of Science Advances. Quantum entanglement occurs when two particles become inextricably linked, and whatever happens to one immediate affects the other – regardless of how far apart they are. Albert Einstein himself described this phenomenon as 'spooky action at a distance.'

We could not properly reproduce these pictures in this book, however when you bring up an internet search for *first-ever photo of quantum entanglement*, you will see directly for yourself these incredible photos. You can well decide for yourself the merit of Albert Einstein's concession that there may exist an unknown, "hidden" variable that acts as a messenger between the pair of entangled particles, keeping their fates entwined.

What new possibilities exist within the idea that entanglement messages can travel faster than the speed of light via an ability to tunnel through barriers, rather than detouring around them? Just as enzymes are a catalyst to accelerate chemical transformations, a better understanding of entanglement and its applications will help us with the tools of tunneling, morphing, and moving through energy barriers.

The first author of the Journal paper mentions the 2019 images as 'an elegant demonstration of a fundamental property of nature'. My own very first thought when I saw these pictures was "Oh wow, these look like the Yin/Yang symbol...what a beautiful correlation!"

~All is connected, all is impacted, all is related~

You may find it helpful at this point to circle back to the earlier mention (under *Setting the Stage*) of the Universal Laws outlined in the Kybalion - and spend some time reviewing, questioning, combining, testing, exploring all those ever-consistent principles.

All new creations that meet with long-term success contain components which coordinate, support, and fulfill those timeless Laws.

"Everything should be made as simple as possible, but no simpler." Albert Einstein 1933

Personal Quantum Energy Application

To move into the action step portion of this primer, here is a short article I wrote several years ago to help set the stage for developing your own personal, unique PQE contributions:

Harmonic Alchemy – Fire and Ice (2016)

Many years ago, I sang with a large choral group directed by a well-known and beloved conductor. The chorale performed its concerts with a major professional orchestra, so there was a great deal of listening, awareness, and intuitive interaction that needed to be present between nearly 200 participants to offer a top-notch performance that was filled with passion and inspiration – yet also balanced timing and coordination. Our conductor often reinforced rehearsals with the analogy, *"Heart on fire, Head on ice"* to help us understand, and practice, this process.

Years later, as I contemplated the idea of spiritual alchemy, this analogy focused my explorations on the fundamental keys of alchemy itself...and it helpfully offers a straightforward understanding of how we all can achieve new creations within ourselves and for our world.

Physical alchemy is the process by which various known elemental ingredients are combined in a dedicated process to produce a desired new element, which is highly sought and prized by both the hands-on creator and others desiring to acquire such value. *Harmonic alchemy* is the process by which each individual soul melds its own current space and energy with new focused intentions in order to continually expand awareness, consciousness, and contributions to its own development and to the universal community.

History is filled with stories of many seeking to use the physical alchemy process to create gold and other elements designated by mankind as 'precious.' And those stories themselves are littered with dire outcomes for those so obsessed with the end goal of accumulating such objects that they disregarded all forms of balance, perspective, and reasoning in their pursuits. There's also been no shortage of stories of those pursuing spiritual transformation who ended up confusing the personal, Source connective, expanding characteristics of pure spiritual alchemy with an over-zealous, fully ego-driven, controlling modality.

In all alchemy explorations, both heat (fire) and cooling (ice) are necessary parts of the creating process. The hottest flame and the coldest ice have the same blue/white color spectrum – just at opposite ends of the temperature scale. The flame provides the opportunity to melt and flow elements together while the crystallization provides the resulting transparent structure and durability of the new creation. And the cycle of heat, add, mix, and cool can continue indefinitely to produce ever more unique, strong, clear creations.

Throughout the existence of Gaia Mother Earth, those using and preserving the healing energy of Source connection have alternated between the hot climates of sea level and the cold atmosphere of the highest mountains to perform their physical and spiritual alchemies...again, both extremes being required to affect transformations to re-remember, re-connect, and re-preserve precious energies otherwise misplaced during human evolutions and experiences.

From a spiritual application perspective, as our <u>heart</u> energies become progressively more open and engaged and our <u>mind</u> energies become progressively clear and insightful, then our personal alchemy experience manifests continually in new states of being having greater vision, self-development, and universal contribution...the concept of the <u>fundamental consciousness element</u>.

There are numerous methodologies and modalities for both heart and mind awareness from all areas of the globe. Whichever ones you authentically choose to assist you will all lead to the same level and quality of experience as another individual's path. There is no need to concern ourselves with comparing whose path 'appears' more dedicated, rewarding, effective, or valued in general human sight – as each is truly equally valid and appropriate.

And our physical body, which provides the actual support and action for and throughout our spiritual experiences, deserves the same deep recognition, gratitude, and healthy attention. *The sum of the parts together is a far greater vehicle for overall manifestation than focus on any one component.*

To assist our awareness focus, here are key
characteristics for harmonic alchemy ~

Heart on Fire:

- ❖ Receptivity
- ❖ Engagement of Vibration
- ❖ Trust
- ❖ Energetic balance
- ❖ Resilience
- ❖ Persistence of Message
- ❖ Mutual connection

Head on Ice:

- ❖ Observation
- ❖ Non-judgment
- ❖ Clarity of Intention
- ❖ Conscious choices
- ❖ Grounding
- ❖ Mutual respect
- ❖ Big picture perspective

~~~~~~~~~~~~~~~~~~~~~~~~~~~~~

So, what will be the soundtrack for *your* life – literally and
figuratively?

Thesaurus.com defines "Navigating Destiny" as: Intention, Expectation, Future, Prospect, Design. These concepts are quite the opposite of the frequently misdirected pronouncement that equates destiny with a pre-determined fate outlook...an illusion of limitation, karmic retribution, outside power, and fear of personal value.

The application of your personal energy is unlimited for both your personal fulfillment and for your service to others. There need be no sense of duality between self and service, and no separation between life on any one planet and the Universe in totality. Latch onto the vibrational songs that have special meaning for you, and sing/play/create your own compositions!

Next Steps and Summary

I imagine Albert Einstein himself would have much to add and clarify regarding this $E=mc^8$ discussion. And, at the same time, encourage all of us to take these ideas being offered and make them our own unique experiments and experience generators. Many have taken the $E=mc^2$ equation and body of information to create applications...now it's our opportunity to move ahead with an expanded body of conscious knowledge and focus.

There is a centuries old proverb that provides an important caution for all us in the Personal Quantum Energy adventure, namely "The road to hell is paved with good intentions." Our 'good intentions', when activated in an ill-prepared, sloppy, or unconsidered manner, can indeed lead to outcomes with unintended consequences (similar to the Ripple Tank effects) from seemingly minimal deviations to catastrophic repercussions.

In the United States, for example, the Model Penal Code (1962) provides a foundational law definition of levels of responsibility for "Actus rea" – the end outcome of an action. Conclusions as to level of harm moves up the scale beginning with "Mens rea" or Intention:

1. The purpose of intention

2. The knowledge of outcome

3. The recklessness of action

4. The negligence of implementation

Whether through a human measuring tool, such as the legal code above, or with a Universal measuring tool, such as the Hermetic Principles, none of us are exempt from the concepts of personal responsibility, appropriate preparation, and continual reassessment and evaluation.

As you draft your potential creations, it's critical to review all the equation factors as integral elements – in as self-honest and highest and best purpose focus as you can at that point in your journey.

The concept of increasing finesse is key to a variety of inputs: self-education and confidence, better targeted outcomes, truly understanding that creation is an evolving cycle, and helping us feel secure in equally valuing our own unique perspectives and thoughtfully collaborating with others (both seen and unseen).

Regardless of your profession or personal identification with specific philosophies, there are opportunities for each of us to tap into the tools and resources (some shared in this primer) to imagine, experiment, and create.

Next Steps

Set aside this primer and literally sleep on the ideas, for as many hours or nights as seems right for you

Set aside all phones, watches, computers, tablets, and similar for regular, dedicated contemplation

Commit to uninterrupted time with a combination of the following:

- Your whiteboard and markers

- Meditation in whatever form suits you

- A retreat into Nature

- A small circle brainstorming gathering

- Immersion into visual art, music, dance

- A fresh new sketch pad or writing journal

- Random tinkering with your lab kit, new materials, a different expression outlet, etc.

- Other deliberate settings for quiet, listening, hearing, receiving, self-opening opportunities

Listen to, and record, all new ideas, visions, pictures, signs, symbols, synchronicities, coincidences, and new appearances in your life – without dismissing, minimalizing, or judging anything

Re-check your recordings for biases toward limitations, 'reasonability', precedents, and the like...and to the best of your current ability Remove/Replace/Upgrade your observations and interpretations

Step back, literally or figuratively as needed, to view the 10,000 ft level repetitions of priorities, hearts desires, burning ideas, and gut level internal purposes that have been made apparent in your contemplations

Regardless of the level of 'craziness' that may seem inherent in those ideas, embrace the new reflections of your consciousness and allow the details and timing of actions to be suspended until the ideas can more fully germinate and grow

Go back to the portions of the primer that seem most significant to you, especially the Finesse Factors and Layers, and consider how incorporating additional tools, skills, and perspectives into your repertoire can leapfrog your past and current ideas into meaningful CREATIONS

Make a commitment to your new creation goals and move along the implementation stages as inspired, understanding that there may well be pauses, re-routings, adjustments, tweaking, tinkering, re-theorizing, and more as your experiments progress.

Relax, repeat, and re-create in infinite cycles!

To Summarize

➢ You embody a tremendous personal power which is a direct portion of the Universal Source energy

➢ You have the free will choice whether to acknowledge this internal power and become an active participant in creating

➢ You have the opportunity to select what tools and resources are appropriate matches to your intentions and purposes

➢ You are capable and empowered to develop any and all areas of importance to you, including: physical health, spiritual exploration, tangible matter products, intangible energetic contributions, principles and vision for business/governing, environmental partnerships, artistic expressions and reflections, perspective-enhancing endeavors, community-building strategies, individual affirmation and encouragement, medical collaborations, and more

➢ You continue your own personal expansion through exploring the Finesse Factors and Layers, and adopting an ongoing infinite creation cycle of question to answers to question to answers...

➢ Your involvement and expression of personal power has an immediate and direct impact on the global and universal energies and outcomes

➢ Your application of global and universal energies, likewise, has a corresponding immediate and direct impact for yourself and the human experience

➢ You get to choose how you experience and respond to energy generated around you from differing sources and intentions

➢ Your combined awareness, empowerment, and action phases determine the sum output of your soul life – at present and through infinity

➢ You are responsible for addressing and eliminating the illusions of limitations in your *own* life. You may offer to assist others *if requested*; if not, it's their personal soul business to pursue – not yours to force.

> ➢ Your experiments and creations that include consideration of all the components of the <u>Energy by Intention theory</u> contribute greatly to your own expanding journey *and* resonate with gratitude throughout the Universal experience

So, from one Energy Explorer to another:

═══

Embrace your personal energetic power, with wonder and joy, as part of the Universe,

Dive into the study, contemplation, and finessed preparation of those ideas that resonate with creativity of your soul, and

May the Force be with You – and in You!

Alberta Einstein, January 2021

Resources and Reference List

Books

Hermetic Principles/The Kybalion by The Three Initiates published 1908

The World as I See It/Out of My Later Years – Albert Einstein, 1990 published compilation

Musical Acoustics (An Introduction) by Donald E Hall, 1980

The Tao of Physics: An Exploration of the Parallels between Modern Physics and Eastern Mysticism by Fritjof Capra, 1975

Space-time and Beyond by Bob Toben, 1975

Quadrivium: The Four Classical Liberal Arts of Number, Geometry, Music, & Cosmology, by Lundy, Ashton, Martineau, and Sutton, 2010

The Complete Book of Numerology – Discovering the Inner Self by David A. Phillips, 1992

The Subtle Body: An Encyclopedia of Your Energetic Anatomy by Cyndi Dale, 2009

What Is Lightbody? by Tashira Tachi-ren, 1999

The Second Coming of Christ by Paramahansa Yogananda, 2004

Your Energy in Action! Energy Balancing for Daily Living by Kabir Jaffe, Ritama Davidson 2013

A Treatise on Cosmic Fire by Alice A Bailey, 1925

Soul Speak: The Language of Your Body, by Julia Cannon, 2013

A Short History of Nearly Everything by Bill Bryson, 2003

The Body: A Guide for Occupants by Bill Bryson, 2019

Awaken to Ascension by Marsha Hankins, 2021

Articles

The Legacy of E=mc^2, Article by Peter Tyson, October 14, 2005 (PBS/NOVA)

Einstein's Theory of Special Relativity, Article by Elizabeth Howell, March 30, 2017 (Space.com)

Albert Einstein and E=mc^2, Article by Deborah Byrd, September 27, 2018 (Human World/EarthSky.org)

Stuff You Should Know Podcast (Stitcher.com) January 13, 2015 "How the Scientific Method Works"

How the Scientific Method Works, Article by William Harris (Science.HowStuffWorks.com)

Scientists Just Unveiled the First-ever Photo of Quantum Entanglement, Article by Fiona MacDonald, July 13, 2019 (ScienceAlert.com)

18 Times Quantum Particles Blew Our Minds in 2018 by Rafi Letzter, December 27, 2018

Video

The Secrets of Quantum Physics, Videos 2016 with Dr Jim Al-Khalili, Surrey University

Albert Einstein | The Mystery of Quantum Physics | Universal Documentaries HD 2016

Quantum Entanglement and the Great Bohr-Einstein Debate | Space Time | PBS Digital Studios, 2016

Stephen Hawking: (2016) Genius – on PBS.org

*While Alberta is always exploring throughout
the Universe, she may be reached at*
ExplorerAEinstein@gmail.com
with your questions and suggestions.

Thank you ~ and wishing you awesome travels!

www.ingramcontent.com/pod-product-compliance
Lightning Source LLC
Chambersburg PA
CBHW070946210326
41520CB00021B/7071